Careers in Forensics™

Careers in Fingerprint and Trace Analysis

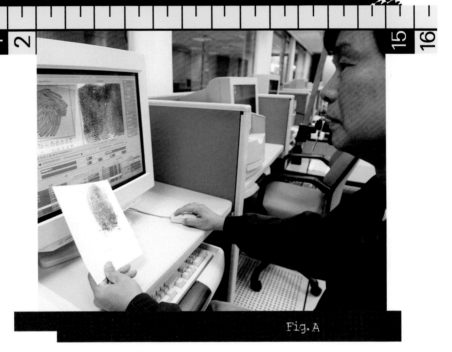

Fig. A

Jeffrey Spaulding

New York

Published in 2008 by The Rosen Publishing Group, Inc.
29 East 21st Street, New York, NY 10010

Copyright © 2008 by The Rosen Publishing Group, Inc.

First Edition

All rights reserved. No part of this book may be reproduced in any form without permission in writing from the publisher, except by a reviewer.

Library of Congress Cataloging-in-Publication Data

Spaulding, Jeffrey.
Careers in Fingerprint and trace analysis / Jeffrey Spaulding.—1st ed.
    p.cm.—(Careers in forensics)
Includes bibliographical references and index.
ISBN-13: 978-1-4042-1344-9 (library binding)
1. Forensic sciences. 2. Fingerprints. 3. Trace analysis. 4. Evidence, Criminal. 5. Criminal investigation. I. Title.
HV8073.S62863 2008
363.25'8—dc22

2007033469

Manufactured in the United States of America

Contents

	Introduction 4
Chapter 1	Catching Criminals 9
Chapter 2	On the Job 21
Chapter 3	Learning the Ropes 32
Chapter 4	Employment 39
Chapter 5	What the Future Holds 46
	Glossary 55
	For More Information 56
	For Further Reading 58
	Bibliography 59
	Index 62

Introduction

Western culture is fascinated by the art and science of evidence detection and analysis and the solving of crimes. In the movies, on television, in books, and even in newspaper stories, we are surrounded by stories of criminal mysteries, investigations, and conclusions.

Since the discovery of fingerprint evidence in the nineteenth century, crime-fiction writers have eagerly incorporated fingerprint and other trace evidence into their plots. In the late 1880s, the famous American writer Mark Twain wrote about a killer being caught thanks to fingerprint evidence in his book *Life on the Mississippi*. Around the same time, a British author named Arthur Conan Doyle began writing a series of stories involving an amateur detective named Sherlock Holmes. Holmes used a revolutionary new crime-solving method: he carefully examined the evidence found at a crime scene and used his powers of deduction to solve the mystery. Dozens of movies have been made about Holmes, who is still one of detective fiction's most popular and enduring characters.

INTRODUCTION

In recent years, fingerprint and trace evidence have begun playing a bigger and bigger role in tales of crime and punishment. Television shows such as *CSI: Crime Scene Investigation*, *Cold Case*, *Forensic Files*, and *Law & Order* have made forensic science more than just a key part of criminal investigation—it's now a pop culture phenomenon. The public is more curious than ever about forensic science.

Forensic science, or forensics, is the application of scientific methodology to the gathering and analysis of evidence. There are a number of forensic sciences, such as DNA analysis, ballistics, and fingerprint and trace analysis. Using advanced technology to gather and analyze evidence, forensic scientists can make startling discoveries.

Television shows such as *CSI: Miami* have turned forensic crime-solving techniques into a popular phenomenon.

Wherever you go, you leave small traces of yourself behind: fingerprints, eyelashes, footprints, skin tissue, and clothing fibers. These traces reveal far more information than meets the eye. A fingerprint may eventually help a forensic scientist identify a criminal, but it can also point to other facts, such as whether the suspect is a smoker or has been handling chemicals. A human hair can be matched with the hair of a suspect but can also yield additional useful information, such as whether or not its owner had certain childhood diseases.

Fingerprint analysis is one of the oldest, and most important, of the forensic sciences. The tips of every person's fingers contain small ridges called fingerprints. No two fingerprints are the same. We now know that fingerprinting was used to identify people in China more than one thousand years ago. But in the Western world, it was not until 1880 that Dr. Henry Faulds published an article suggesting fingerprints could be used to identify people. In 1892, Sir Francis Galton published his book *Finger Prints*. That same year, in Argentina, the first-ever criminal conviction based on fingerprint evidence was handed down. Five years later, in Calcutta, India, a bureau was established to collect fingerprints for the purpose of building criminal fingerprint records. In 1901, a fingerprint bureau was established at Scotland Yard, the main police investigative organization in England (akin to the Federal Bureau of Investigation, or FBI, in the United States). It wasn't long before fingerprinting became a widely accepted technique by criminal justice organizations the world over.

But fingerprints are not the only evidence that criminals leave behind—a fact recognized by the father of forensic science, French scientist Edmond Locard (1877–1966). Locard was responsible for establishing the first-ever crime laboratory. The Locard Exchange

Edmond Locard was one of the most important figures in the history of crime fighting. Locard dedicated his life to the discovery and perfection of forensic science techniques.

Principle is the cornerstone of forensic science: "Wherever he steps, whatever he touches, whatever he leaves, even unconsciously, will serve as a silent witness against him."

Today, trace evidence can be nearly anything left behind at a crime scene, such as a strand of hair, a piece of rope, clothing fibers, paint chips, mud or dirt, or gunshot residue. The analysis of this evidence can help link a suspect to a crime scene. For instance, textile fibers might be matched to an article of the suspect's clothing, or a piece of rope can be matched to a particular manufacturer. By itself, trace evidence is usually not enough to determine a suspect's guilt or innocence. But in conjunction with other data, such as fingerprint matches, evidence provided by trace analysis can be a decisive factor in the outcome of a criminal case.

According to Locard, forensic evidence is the ultimate witness. If properly gathered and properly analyzed, fingerprint and trace evidence can provide crime fighters with invaluable knowledge. Forensic evidence can provide the necessary proof for a jury to convict a criminal. On the other hand, forensic evidence can also help clear an innocent suspect of wrongful charges.

Forensics is more than just a fascinating way to catch criminals. It's a science that is more vital than ever and a career with limitless future potential. Forensic scientists will continue to play an important role in criminal justice in the foreseeable future.

Chapter 1
Catching Criminals

The first case ever solved through the use of fingerprint analysis was a violent robbery that occurred in Deptford, England, in 1905. Two burglars robbed a store and, in the process, brutally murdered the couple who owned it. While the case might not seem remarkable by today's standards, it was a landmark in the history of forensic science.

Because the analysis of fingerprint evidence was still in its infancy and an unproven technique, Scotland Yard had been hesitant to use it in criminal cases. But after finally becoming convinced of the validity of fingerprint evidence, it established a fingerprint bureau in 1901.

Police investigators discovered a wealth of circumstantial evidence at the crime scene, but nothing definite enough to link the suspects to the crime. Luckily, a fingerprint was discovered at the scene. It was a thumbprint, to be precise, and it was found on a cash box that had been opened. The thumbprint was compared against the thumbprint of suspect Alfred Stratton. When the two prints were found to match, Stratton was sent to the gallows and hanged.

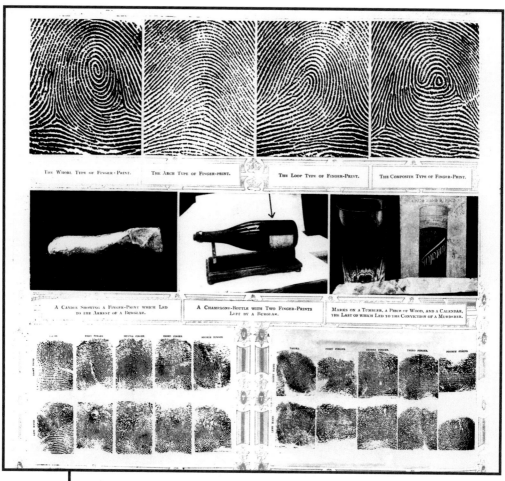

This Scotland Yard fingerprinting guide dates from around the time of the first-ever use of fingerprint evidence in a criminal trial. Scotland Yard's fingerprint bureau would prove to be a landmark in forensic science.

Interestingly enough, the first-ever fingerprint case was not without its controversies. Dr. Henry Faulds did not think the thumbprint on the cash box was clear enough to provide a match with Stratton. In fact, he went as far as to serve as a witness for the defense. Still, the success of fingerprinting in this case legitimized it in the eyes of the law enforcement community. Indeed, throughout the years, fingerprint analysis has been behind the capture of some of America's most feared criminals.

Public Enemy #1

It wasn't long before the U.S. law enforcement community began fingerprinting all criminals upon their arrest. Besides simply worrying about getting caught in the act, criminals now had to worry about leaving prints behind. Once criminals had committed one crime and been arrested for it, their prints would remain on file for the rest of their lives, potentially linking them to any crimes that they might commit in the future.

John Dillinger was a career criminal who would eventually become uncommonly concerned about fingerprints. Born in Indiana in 1903, Dillinger fell into the criminal life while still a young man. After getting caught robbing a grocery store, Dillinger was given a harsh sentence. He spent nearly a decade in prison and emerged an angry man determined to break the law once again.

It was 1933, and the Great Depression had stricken America. The U.S. economy hit a record low, and millions upon millions of people were out of work. Government programs created work for many, but there was still not enough money to go around. Many people were poor, hungry, and desperate.

Shortly after leaving prison, John Dillinger attempted to hold up a bank. He was quickly apprehended, sentenced, and put back

This poster of John Dillinger circulated during the last year of his life. Dillinger became instantly recognizable, which ultimately worked against him.

in jail. Soon, he was broken out of prison by a group of gun-wielding friends. Once free, Dillinger and his gang went on a whirlwind crime spree that quickly vaulted them into the ranks of America's most wanted criminals.

After a subsequent arrest, Dillinger again escaped from prison with the help of a fake gun that he had carved out of wood. The gutsy criminal stole a policeman's car and drove it across state lines, drawing the attention of the FBI. Slowly but surely, the law began closing in once more.

Trying to lie low, Dillinger found that he was now strangely popular with average, struggling Americans. Murderous criminals like John Dillinger, who simply demanded cash from bankers at gunpoint, were becoming unlikely folk heroes, especially to Depression-era Americans whose houses or farms had been repossessed by banks for nonpayment of mortgages or loans. The FBI had declared him "Public Enemy #1," making him

The Permanent Mark

John Dillinger was not the only criminal who tried to remove his fingerprints. No matter what ingenious methods criminals come up with to remove their prints, they are almost always met with failure. The skin on a person's fingers, palms, and soles of his or her feet has friction ridges. These ridges help you grab and hold things—they literally create friction. Fingerprints are friction ridges, and a person's friction ridges always grow back in exactly the same pattern. In fact, these ridges were present before you were born and do not change as you age. There are some criminals, like John Dillinger, who have gone to great lengths to get rid of their fingerprint ridges, resulting in nothing more than mild scars to their fingertips. Ironically, these scars only made the criminals' fingerprints easier to identify.

It is possible to remove fingerprints. They must be surgically removed and replaced with skin harvested from another part of the body. Still, this wouldn't do a criminal very much good. Criminals can still be identified by prints taken from their palms and even the soles of their feet, not to mention whatever trace evidence they might leave behind.

the first criminal to earn the title. His photograph was on wanted posters, and his face was familiar to the average person on the street. How could Dillinger escape when he was recognized everywhere he went? In an attempt to draw less attention to himself, Dillinger underwent black market plastic surgery.

There was just one problem—even if he changed his face, Dillinger's fingerprints (and the prints of many men in his gang) were on file. He could move far away and he could surgically alter his face, but his fingerprints would always give him away. Dillinger was so worried about fingerprint evidence that he attempted to burn his fingerprints off with acid. This was an incredibly painful procedure, and when it was over, his fingerprints had been utterly destroyed. Or so it appeared—after his fingers healed, Dillinger was dismayed to find that his fingerprints had grown back and were exactly the same as they had been before!

Dillinger's ruse didn't work. Despite his plastic surgery, people still recognized him and it was impossible for him to change his fingerprints. Try as he might, John Dillinger couldn't avoid having to pay for his crimes. It wasn't long before FBI agents tracked him down and shot him dead.

Catching the "Night Stalker"

As time went on, the U.S. law enforcement community developed an extensive fingerprint collection. Even before computers helped coordinate print identification efforts, fingerprint evidence managed to bring down a number of very famous criminals. One of the most notorious of these criminals was Richard Ramirez, who would become known as the "Night Stalker."

In 1984, Los Angeles, California, was terrorized by a series of brutal crimes. Someone was breaking into homes, attacking residents, and robbing them. In June 1984, the criminal killed his first victim. It appeared to be an isolated murder, but after a few months, the killer struck again. This murder initiated a string of violent assaults and murders that had law enforcement agents scrambling to find a pattern to the crimes. Witness descriptions of the murderer did not provide enough information to positively identify him. The media dubbed him the Night Stalker.

If not for fingerprint evidence, the Night Stalker might never have been caught. Unlike many serial killers, he seemed to have no predictable pattern; it was impossible to tell where he would strike next. However, he did leave trace evidence behind at some of the crime scenes. The police made casts of his footprints and noted that he drew pentagrams at many of the locations.

The killer's downfall came when a witness saw him driving away from the scene of a crime. His license plate number was traced to a stolen car, which the police were able to locate. Police were also able to recover latent prints from the car, which were matched with Richard Ramirez, a young man with a number of prior arrests. It wasn't long before Ramirez's face was plastered on wanted posters

During his 1985 Los Angeles murder trial, Richard Ramirez displays a pentagram he'd drawn on his hand. Ramirez is currently on Death Row awaiting execution.

and newspapers all around town, throughout the state, and nationwide. There was nowhere he could go without being recognized. Ramirez was quickly caught and convicted. One of the highest-profile criminal cases in American history was brought to an end.

Modern Justice, Modern Solutions

Criminals have long been afraid of inadvertently leaving fingerprints or other trace evidence at a crime scene. Today, modern technology has made it easier than ever for forensic scientists to extract accurate information from trace evidence and to match fingerprints to a suspect.

In 1986, a professor at the University of Michigan–Flint named Margarette Eby was murdered, and the crime went unsolved. Five years later, in 1991, flight attendant Nancy Ludwig was killed in a hotel room in Michigan, and that crime was also unsolved.

But by 2001, a national fingerprint database had been established, allowing law enforcement agents to compare a fingerprint found at one crime scene with a fingerprint taken from any other crime scene throughout the country. This system, known as the Integrated Automatic Fingerprint Identification System (IAFIS), contains fingerprint records on more than forty-seven million individuals.

Using this new database, law enforcement officials were able to link DNA evidence from the two Michigan crimes. A fingerprint left at the Eby crime scene was identified as belonging to a man named Jeffrey Gorton. A DNA sample taken from Gorton was matched to DNA evidence left at the other crime scene as well. Years after both crimes had been committed, fingerprint and trace evidence allowed the killer to finally be brought to justice.

Margin of Error

It should be noted that fingerprint evidence is not foolproof. Convictions are often not gained on the basis of fingerprint evidence alone, but rather in combination with other forensic evidence gathered at the scene of a crime. And despite the fact that no two fingerprints are the same, fingerprint analysis is not entirely free of human error.

No one knows how many people each year are wrongly convicted of crimes due to fingerprint evidence. Many believe false fingerprint convictions to be extremely rare, although there is no way to know for sure. There have been cases of people who have been convicted of a crime and sent to prison, only to be exonerated (cleared of guilt) years later thanks to evidence gathered from different forensic methods, such as DNA analysis.

The Case of the Century

Trace evidence is commonly analyzed alongside fingerprint evidence. Sometimes, trace evidence can provide key information affecting the outcome of a case, but sometimes it simply isn't enough.

One case in which a veritable mountain of trace evidence did not lead to a conviction was the 1995 murder trial of O. J. Simpson. Simpson, a broadcaster and former professional football player, was a beloved American celebrity. In June 1994, Simpson's ex-wife, Nicole Brown Simpson, was found stabbed to death. A man named Ronald Goldman, an acquaintance of Nicole Brown Simpson's, was also found stabbed to death. Simpson was captured by police after a long, low-speed chase through Los Angeles. The live coverage of

Marcia Clark, one of the lawyers for the prosecution in the 1995 trial of O. J. Simpson, presents forensic evidence to the judge and jury. Due to irregularities in the way the evidence was collected from the crime scene, the jury would be led to doubt its validity.

Simpson evading police officers in his white Ford Bronco became one of the most sensational television moments of the 1990s.

The prosecution had a very strong case against Simpson. Trace evidence gathered from the crime scene included blood that was matched to blood inside Simpson's Bronco, his driveway, and his home. It included some blood confirmed to be Simpson's, and Simpson himself had sustained a cut on his hand. Hair found at the crime scene was matched to Simpson's. Investigators also recovered fibers at the crime scene that were matched to fibers in the floor carpet of Simpson's Bronco.

Other trace evidence included a bloody glove left at the crime scene that was paired with a glove found at Simpson's home, and shoe prints at the crime scene were matched to prints found inside Simpson's Bronco. The prints were of the same size and unique luxury brand shoe that Simpson wore.

This trace evidence, combined with other evidence in the case, such as witness testimony, was not invincible. The defense team focused on the reasonable doubt that existed in the case—there was a chance that the evidence may have been tampered with. Specifically, doubt was cast upon the credibility of Los Angeles Police Department (LAPD) detective Mark Fuhrman, who led the investigation. Ultimately, trace evidence was unable to either convict Simpson or convincingly point to a new suspect.

Chapter 2
On the Job

Good forensic science depends on teamwork. Crime scene investigators gather evidence from the scene of the crime. Sometimes, latent print examiners might be called in to collect fingerprints. In a laboratory, forensic scientists examine the evidence gathered by crime scene investigators and print examiners. Their findings may be used by law enforcement officials to solve a crime or by prosecutors attempting to persuade a jury of a defendant's guilt.

Arriving at a Crime Scene

Analysis of fingerprint and trace evidence can yield incredibly important information. However, if collected improperly, evidence can become compromised, contaminated, and useless. Crime scene investigators, or CSIs, must follow detailed procedures in order to ensure that evidence is not contaminated.

After arriving at a crime scene, the CSI's first task is to establish what type of crime took place. Was it a

A crime scenes is marked off with police tape. This ensures that no one accidentally enters the area and disturbs any of the evidence.

robbery, a murder, or something else? The CSI determines the size of the crime scene and secures it. He or she marks off the crime scene with police tape or barricades in order to make sure that onlookers and journalists do not accidentally disturb anything. Crime scenes can change with time. Cautious CSIs might mark off an area larger than the actual crime scene at first in order to protect evidence that may be found at a distance from where the crime occurred. The area of a crime scene can be reduced once it is marked off, but it cannot be easily or effectively enlarged after the initial establishing of a perimeter. The CSI takes careful note of every single detail at the crime scene. If any specialists, such as latent print examiners, are needed to analyze the scene, the CSI calls them in.

Once inside the perimeter of a crime scene, police and investigators take care not to disturb evidence. Fingerprint and trace evidence can easily become contaminated, rendering it inadmissible in a court of law. The location of evidence is important, and evidence like hairs and fibers can be easily moved around a crime scene by a careless investigator. Latent fingerprints can be accidentally wiped away.

Being a CSI can be a tough job. The hours are long and are sometimes irregular. CSIs may have to get up in the middle of the night to go to the scene of a crime. But no matter how tired they are, they must remember that the work they do will affect the outcome of a trial.

Collecting Trace Evidence

Trace evidence is often collected before CSIs get to work collecting fingerprints. Every piece of evidence is photographed before it is removed. Larger pieces of evidence, such as murder weapons, articles of clothing, and pieces of glass, may be removed before smaller pieces of evidence, such as hairs or fibers.

Investigators must take elaborate precautions to ensure that they do not accidentally transfer hair or fiber evidence to other locations within the crime scene. They also usually make sure that each item removed from the scene is placed in its own separate bag. If there is a victim at the crime scene, special care must be taken to keep items belonging to the victim separate from items thought to belong to the suspect.

To avoid any cross-contamination, different people usually collect and bag evidence thought to be left by the suspect and the victim. One investigator's job is to keep a log, or detailed list, of the evidence that other investigators collect. Once bagged and recorded, trace evidence is sent to a laboratory to be analyzed by an examiner.

Examining Trace Evidence

In real life, CSIs don't interview suspects or people who may have witnessed the crime. They simply collect and record evidence. Most subsequent analysis of evidence occurs in laboratories.

In the laboratory, the examiner goes over the evidence. He or she makes close inspections of hats, shoes, socks, or other items of clothing taken from the crime scene. Clothing often yields hair or fiber evidence. Based on what he or she knows about the particulars of the case, the examiner begins the painstaking process of sifting through the evidence.

Hair Evidence

Hairs are viewed with a high-powered microscope. This can tell the examiner what part of the body the hair has come from and whether it was naturally shed or forcibly removed. It is usually possible to tell the race of the person to whom the hair belonged. Sometimes, the examiner can also tell whether the person was a male or female.

ON THE JOB

Hair evidence can provide law enforcement agents with a surprising amount of information. Here, a lab technician conducts a test to see if this hair has been chemically dyed.

When trying to match a hair recovered from a crime scene to, say, a criminal suspect, the examiner compares it to a sample taken from that person. Under the microscope, the examiner determines whether or not the hairs match. Matching a hair sample to a suspect is not the same as matching a fingerprint to a suspect, however. While it is rare for two people to have hair that matches under microscopic analysis, it is possible. Hair evidence can be very valuable, but it needs to be considered alongside other, more definitive evidence.

Fiber Evidence

Much like hair evidence, fiber evidence may or may not have any importance or validity, depending on the particulars of the case. In the laboratory, the examiner simply analyzes the fiber evidence—it is up to law enforcement officials to determine whether or not the evidence is important by providing some link to a suspect. Fiber can be valuable alongside other evidence.

It is possible for examiners not only to determine whether the fiber is natural (such as cotton, wool, or silk) or synthetic (such as polyester, nylon, or rayon), but also what company manufactured the fiber. A fiber taken from a crime scene, for instance, could be matched with a fiber from a suspect's nylon jacket. Or a fiber retrieved from a victim's shoe might match nylon carpet fibers found in a suspect's car.

Dust, Glass, and Paint Evidence

Dust and glass at a crime scene can provide valuable evidence. There are different kinds of dirt and dust, and sometimes, laboratory analysis can determine where they come from. Pieces of broken glass at a crime scene might mean there was a struggle or that someone broke into a residence. Paint evidence can be useful, too, and laboratory analysis can match a chip or flake of paint found at a crime scene to the paint on a suspect's car or in his or her home.

Evidence and Analysis

Other trace evidence, such as material recovered from beneath a victim's fingernails, glass fragments, or gunshot residue, is also analyzed in the lab. Sometimes, the examiner will submit certain

ON THE JOB

trace evidence—like hairs discovered on a suspect's clothes or scratched skin found beneath a victim's fingernails—for DNA analysis. If the hair matches that of the victim and is determined to have been pulled out forcibly, this evidence can be used against the suspect. If the DNA of the skin found under the victim's fingernails matches that of the suspect, it would help establish that the victim was killed by the suspect after a violent struggle.

Fingerprint Evidence

After trace evidence is gathered, CSIs or a latent prints unit may go over the crime scene for fingerprints. Fingerprints may be found on any surface at a crime scene, including certain types of fabric, the skin of a deceased person, or a bar of soap. A number of specialized techniques have been developed to document fingerprints, depending on what kind of prints they are.

There are three kinds of fingerprints. Plastic fingerprints are left as impressions in soft material, such as wax. Patent fingerprints are left when one substance is transferred to another, such as a muddy fingerprint on a wall. Latent fingerprints cannot be seen with the naked eye. They are formed by the natural oils and skin cells that are left behind when a person touches a surface.

Plastic and patent prints are often photographed for examination later on. Latent fingerprints are invisible, and a great number of methods exist for collecting and exposing them. These vary due to

An evidence technician searches for fingerprint evidence. By dusting this entryway with powder, he is hoping to expose latent prints, which are not visible to the naked eye.

what kind of surface the prints are found on. For instance, prints can be made visible by dusting them with various types of powder; by exposing them to various dyes, chemicals, or sprays; or by shining ultraviolet (UV) light on them.

With today's technology, fingerprints may reveal a great deal of surprising information. For instance, residue left behind in the print may indicate if the person who left the print is a smoker, has been handling explosives, has fired a gun, or has been using drugs.

ON THE JOB

Points of Comparison

Even with the availability of comprehensive databases like IAFIS, a comparison between two fingerprints is still verified by a human being. When comparing prints against each other, the print examiner looks at them very closely, trying to match distinctive shared characteristics.

Fingerprints are generally defined by one of three basic patterns that they might have. These patterns are loops, whorls, and arches. Each print features unique, minute details, such as whether individual ridges end, split, or cross. These minute print details are usually specific to individual fingerprints and are used as points of comparison when print examiners are trying to match prints from a crime scene to prints on file. Generally speaking, twelve points of comparison have to be found between two fingerprints before they can be said to be identical.

Careers in Fingerprint and Trace Analysis

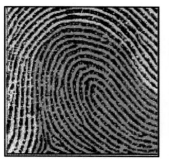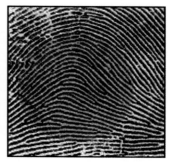

No two fingerprints are alike. However, all share similar characteristics. The most general characteristic of a fingerprint is the arrangement of its ridges, as shown in these images from the FBI Web site.

Evidence and Its Application

During a typical day, a latent print examiner can expect to spend the greater portion of his or her time engaged in collecting and analyzing evidence. Collecting, comparing, analyzing, and verifying fingerprints is a time-consuming job, even with the help of computer databases. Forensic scientists analyzing trace evidence also have an exacting job to do. Examining trace analysis is painstaking, methodical work.

However, latent print examiners, trace evidence analysts, and forensic scientists do more than just sift through evidence. They prepare written reports of their findings for law enforcement officials. These reports put forth everything that they learned about the evidence and explain what the lab analyses mean in comprehensible layperson's terms. Reports do not offer opinions on the case and

A forensic scientist presents his findings during a 2004 trial. Courtroom testimony is one of the most important aspects of a forensic scientist's job.

do not draw conclusions about a suspect's guilt or innocence—that is for the jury to decide.

However, forensic scientists are often called upon to testify during a trial. While giving their testimony, they must present their findings in such a way that the lawyers for the prosecution and the defense, the judge, and the jury can understand them. Outside of the courtroom, forensic scientists and evidence examiners may spend a lot of time talking to attorneys and law enforcement officials on the phone, doing paperwork, and otherwise engaging in typical administrative tasks.

Chapter 3
Learning the Ropes

There is a lot of competition for jobs in forensics. If you are interested in a career in forensics, it is a good idea to start preparing early. Employers, from the FBI to local police departments, look for serious candidates with college degrees and backgrounds in science. Those who manage to secure a job will find that a career in forensics can be both challenging and time-consuming but extremely rewarding.

Getting an Education

If you are interested in a career in fingerprint or trace analysis, you must get a solid science education. Forensic science is extremely specialized, and it is important to have an excellent base of general scientific knowledge to build upon. In high school, get good grades and take classes that will prepare you for college-level science. You might even consider taking college-level science and math courses at a community college while still attending high school. In college, take a wide range of science

A teenager prepares to participate in a demonstration of how police can analyze blood spatter evidence. This hands-on activity is part of an educational CSI program for high school students at a crime laboratory in Albuquerque, New Mexico.

classes. Classes such as chemistry, biology, physics, calculus, and biochemistry will give you the scientific grounding you need for a career in forensics. Besides the various sciences, classes in mathematics, statistics, ethics, criminal justice, writing, and public speaking will also prove invaluable in your career and will be attractive to potential employers.

There are about 100 colleges and universities in the United States offering degrees in forensic science. A forensic science curriculum

These law enforcement personnel are listening to an instructor at the National Forensics Academy. They are examining a car destroyed by fire.

teaches students the basic science they will need before embarking on a career in the field, much like pre-med majors take all the science classes required to get into medical school.

Beyond the Classroom

Forensic scientists spend a lot of time in the laboratory, so it is important to get as much laboratory experience as possible while still in college. Many police agencies across the country offer forensic internships, which will help you build your resume and give you practical experience to draw on throughout your career.

With so much popular interest in forensic science, competition for jobs in the field has become fierce. Beyond a solid education and impressive internships, additional relevant work experience can give you an edge over other job seekers. For instance, many crime scene investigators began their careers as police officers. This is

Internships

Internships are a great way to learn about fingerprint and trace analysis. As an intern, you will be working among professional fingerprint and trace analysts, watching and perhaps assisting them as they do their jobs. Although college can give you the educational foundation you need for a career in this field, it can't entirely prepare you for what it is like to work among professional crime scene investigators and lab analysts. Some internships are unpaid, while others may offer stipends (small allowances). But more important than money, you will get real-world experience in your career of choice and make important contacts in the field of forensics.

The FBI offers the Honors Internship Program, which selects college students to spend ten weeks at FBI headquarters in Washington, D.C., during the summer. Interns are assigned to an FBI division depending on their field of study. This is a very prestigious internship and a great way to gain valuable experience. Applicants must be juniors or seniors in college or graduate students.

a valuable entry point into forensics because police officers work within crime scenes, observe evidence gathering firsthand, and learn what it means to be a part of the law enforcement community. Being a police officer can give you the kind of experience that will eventually help qualify you to be a CSI.

Success in fingerprint and trace analysis involves many other skills that have nothing to do with science or criminal justice. Forensic scientists must have good interpersonal and time-management skills, since they often work alongside others in high-pressure, high-stakes situations while facing strict deadlines. Good writing and speaking skills are also important—if you cannot clearly articulate the nature of your findings, the work you have done is useless and will hamper the prosecution of suspects.

Background Checks

A career in fingerprint or trace analysis, like every career in forensic science, is forged within the criminal justice system. To this end, candidates for jobs in forensics must pass background checks if they hope to be employed. This means that a history of drug use, criminal convictions, a bad credit history, a poor driving record, or links with politically subversive organizations can endanger a candidate's future career prospects. A model candidate for a career in forensics is someone who excels academically, has appropriate work experience, and demonstrates a great amount of integrity in his or her personal life.

This crime scene technician works to develop latent prints on glass. There are different methods for exposing prints on various surfaces.

Training

After getting an entry-level job in criminal forensics, new hires go through a training period. Depending on the workplace, the training period can last anywhere from a few months to several years. On-the-job training and education may extend long after the official training period has finished, however. Forensic science is a rapidly changing field that has progressed by leaps and bounds over the last few decades. Within a few years, current techniques of fingerprint

CAREERS IN FINGERPRINT AND TRACE ANALYSIS

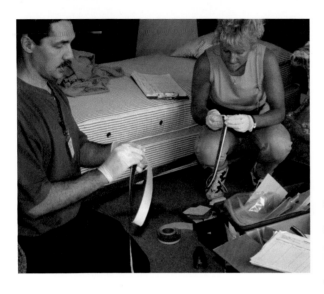

Frequent training helps forensic scientists and law enforcement officials stay current. The FBI agents in this photo are undergoing training for evidence collection at the FBI Academy.

and trace analysis may seem antiquated and be replaced by more technologically advanced methods. So, it is normal for forensic scientists and crime scene investigators to undergo training throughout their careers that will keep them up-to-date on the latest forensic techniques.

The FBI Academy

The FBI Academy is one of the most important law enforcement training centers in the United States. Located in Virginia, it provides training in a number of disciplines, including law, ethics, behavioral science, education and communication, and forensic science. The academy's Forensic Science Research and Training Center offers a number of specialized courses for forensic scientists, laboratory personnel, and law enforcement agents. It offers training in advanced evidence examination techniques, laboratory skills, and basic forensic procedures.

Chapter 4 Employment

There are a great number of careers available to people interested in fingerprint and trace analysis. Whether you want to investigate crime scenes and collect evidence or analyze that evidence in labs, work for a law enforcement agency, serve as an expert witness for prosecuting or defense attorneys, or use your skills within the private sector, there is a job out there for you.

Crime Scene Investigator

Being a crime scene investigator allows you to work alongside police and detectives. You will make sure that emergency response crews and police do not disturb evidence. You will decide how evidence should be collected, what order it should be collected in, and what path investigators and examiners should take through the crime scene during evidence collection to minimize the risk of evidence contamination. You decide whether specialists need to be called in, and you talk with the police about what they saw, smelled, and heard upon first arriving at the crime scene. You will also identify and separate witnesses and suspects, if any are on hand.

Being a CSI is a good way to get hands-on experience about the world outside the laboratory. The hours are long and irregular, but each crime scene offers its own particular challenges, and the job itself is vital to criminal investigations. If evidence is mishandled at a crime scene, it could render the entire case useless and the criminal may go free.

Evidence Collection Toolkit

Crime scene investigators and latent print examiners need special equipment to do their job. For collecting fingerprints, latent print examiners often carry equipment such as brushes, various chemicals and powders, a flashlight, tape for lifting prints, and cards upon which to transfer the prints. Latent print evidence can be very fragile, so latent print examiners and crime scene investigators must take special care not to disturb it.

For collecting trace evidence, crime scene investigators carry glass vials, forceps, tweezers, lifting tape, and a special vacuum cleaner designed for evidence collection. CSIs are in charge of having the evidence collected, documented, packaged, and transported to a laboratory. CSIs must be alert while surveying the crime scene and collecting evidence. Sometimes, criminals try to alter a crime scene to confuse police. Evidence may have

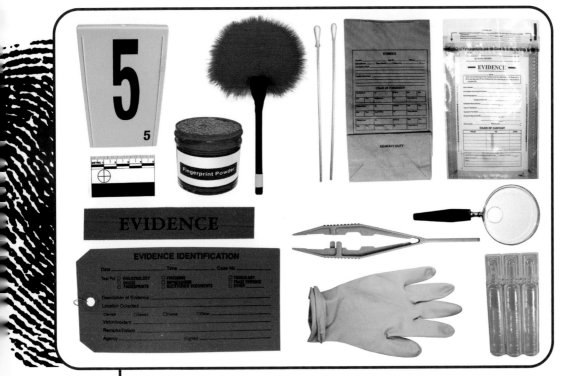

Special equipment is needed for evidence collection. Crime scene investigators may carry rubber gloves, tweezers, a magnifying glass, and powders and brushes to dust for fingerprints.

been moved, doctored, or destroyed. Alert crime scene investigators pay special attention to crime scenes that have been tampered with.

Lab Work

Even with automated fingerprint databases like IAFIS to help match fingerprints, print matching is ultimately done by human beings. When a print or prints are fed into IAFIS for comparison, the database is searched for other prints with similar characteristics. A print examiner looks at the results the database turns up and sets about manually examining the prints to see if they match. The sheer variety of trace evidence that may be collected at a crime scene means that the trace evidence examiner might specialize in analyzing a certain kind of evidence, such as ballistic evidence. Working in a laboratory isn't all science, however. You will have a number of administrative duties as well. You may also provide courtroom testimony on your findings.

Academics

Decades of research into better methods of fingerprint and trace evidence analysis have yielded the advanced forensic science that the American law enforcement community is renowned for today. The FBI's Counterterrorism and Forensic Science Research Unit (CFSRU) works actively with researchers and academics to develop new forensic technologies. As an academic in the field of forensics, you will have ample time to apply yourself to the study of forensic science. The research conducted by academics in fields such as fingerprint and trace analysis can lead to new methods of evidence collection and analysis. This, in turn, can lead to more criminals being brought to justice. Many academics in the field of forensic science publish the results of their research in scholarly journals. Academics can make a career teaching at a college or university and providing expert testimony or analysis for courts and the media.

EMPLOYMENT

Fighting Terrorism

Terrorist activity has posed a unique challenge to the U.S. law enforcement community. Terrorists can be of any race or of any religion, and they can hail from any country. They can even be "home-grown" Americans. Terrorists do not wear uniforms and are indistinguishable from law-abiding members of the civilian population. It is important for law enforcement agencies, on both the local and federal levels, to be aware of possible terrorist activities. Fingerprint and trace analysis can help them link terrorist activity, or suspected terrorist activity, with an individual or group.

The FBI's Evidence Response Team Unit (ERTU) uses sophisticated technology in the process of collecting evidence related to acts of terrorism or evidence that might aid in investigations regarding suspected terrorists. The ERTU is a mobile unit able to travel to the scene of each investigation. The unit is trained to gather any and all forensic evidence, including fingerprint and trace evidence, from an investigation scene under nearly any circumstances.

In Iraq, soldiers in the U.S. Army have begun collecting fingerprint evidence from suspected terrorists and insurgents and comparing it with fingerprint evidence collected from battle sites. The fingerprints are deposited in a central database via satellite. Not only does this evidence help the army identify terrorists and insurgents, it also allows it to verify that native-born Iraqis working for the U.S. government in Iraq do not have a known criminal or politically radical background.

Disaster Work

Not all forensic work is used to fight crime. Sometimes, fingerprint and trace analysis is used to identify people in the event of large disasters. The ERTU has responded to disasters such as Hurricane Katrina, which flooded New Orleans, Louisiana, in 2005, and the tsunami that devastated Southeast Asia in 2004.

Hurricane Katrina took the lives of nearly 2,000 people. ERTU agents managed to fingerprint more than 800 of the hurricane's victims, eventually positively identifying at least 100 of them. The 2004 tsunami in Southeast Asia took the lives of at least 200,000 people. By working with local and international law enforcement and forensic agencies, FBI latent print analysts helped identify a number of tsunami victims.

Private Investigation

Private investigators work for individuals or businesses. They may investigate employee fraud, run background checks, or place surveillance on a particular person. Private investigators are not associated with the police, the military, or the government. They are private individuals who often have a degree in criminal justice or have previously served in some sort of law enforcement capacity.

This French fingerprint specialist searches through prints taken from Thai victims of the 2004 tsunami that devastated large areas of Southeast Asia. A massive effort to identify tsunami victims by their fingerprints was undertaken by local and international law enforcement agencies.

Many private investigators work for companies, many of which specialize in a particular kind of investigation. Investigation into insurance claims and insurance fraud, for instance, may involve cases with millions of dollars on the line. Some larger private investigation companies have their own forensic teams. In many ways, collecting or analyzing fingerprint or trace evidence for a private detective firm is not so different from doing the same job for a police department or even the FBI.

Chapter 5
What the Future Holds

Fingerprint and trace analysis has come a long way since the discovery of the fingerprint. Advances in forensic technology have allowed examiners to locate and recover fingerprints and trace evidence from nearly any surface—including those submerged under water—under any set of circumstances. Today, fingerprints that would have remained invisible decades ago can be revealed thanks to high-tech techniques such as applying silver nitrate to prints or exposing them to infrared light. Hairs left behind at a crime scene can not only be visually compared to the hair of a suspect, they may also be subjected to DNA analysis in order to conclusively prove (or disprove) a genetic match. A clothing fiber, a fingernail, or a bit of gunpowder can reveal an astounding array of facts—and forensic scientists are constantly developing ingenious new ways to extract additional information from the tiniest of crime scene traces.

But the basic principles of fingerprint and trace analysis remain constant. Locard's Exchange Principle—"Wherever he steps, whatever he touches, whatever he leaves, even unconsciously, will serve as a silent witness against him"—

New technologies ensure that forensic science techniques will continually evolve. Here, ultraviolet light is used to expose latent prints that would otherwise be invisible.

still holds true. The future of forensic science lies in two major areas: the development of still more advanced techniques for analyzing evidence and identifying criminals, and the devising of new methods of organizing and retrieving the information this evidence reveals.

Looking Toward the Future

Forensic science and technology will continue to become more accurate, more advanced, and more efficient. However, this does not necessarily mean that there are an unlimited number of jobs in the field of fingerprint and trace analysis. Law enforcement agencies are subject to government funding at the federal, state, and local levels. There are many understaffed and underfunded police departments. Paying salaries to crime scene investigators, forensic scientists, laboratory administrators, and technicians, in addition to having to buy sophisticated forensic technology, can be extremely expensive. Many police departments would like to have large forensic units and cutting-edge crime labs, but they simply cannot afford to.

New crime scene investigators and forensic scientists may always be needed, but unfortunately, government funding will determine the number of forensic scientists that can actually be hired. There are no guarantees. Depending on the amount of funding available, there could be an abundance of jobs available for hardworking men and women—or very few.

New Techniques

New forensic techniques are changing the way law enforcement officials, and forensic scientists, approach fingerprint and trace analysis. For instance, the ability to analyze DNA revolutionized the field of forensics. DNA is the genetic information contained in every cell in your body. Every living being on Earth has a unique DNA code—in fact, scientists sometimes refer to DNA as a "genetic fingerprint." If biological trace evidence, such as dead skin cells, hair, blood, or saliva, is found at a crime scene, the

What the Future Holds

DNA found in these traces may be analyzed. DNA codes are even more elaborate and unique than fingerprints. Therefore, matching a DNA sample from a suspect and a DNA sample found at a crime scene can be extremely effective and definitive evidence. And DNA evidence can free the innocent as often as it helps in capturing the guilty. Since the advent of DNA identification in the 1980s, hundreds of long-standing criminal convictions have been reversed after DNA evidence proved a convict's innocence.

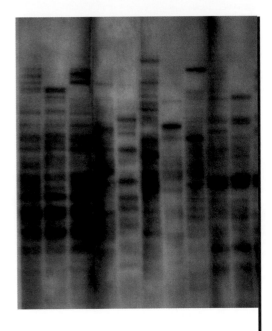

Each person has a unique "genetic fingerprint," such as the one seen here. The ability to match people to DNA samples was developed in the 1980s.

But DNA analysis is not the only type of identification that forensic scientists have developed. The wider field of biometrics is poised to turn the field of forensics upside down.

Biometrics

We are all familiar with the criminal descriptions used to identify perpetrators in movies and television shows—height, weight, and

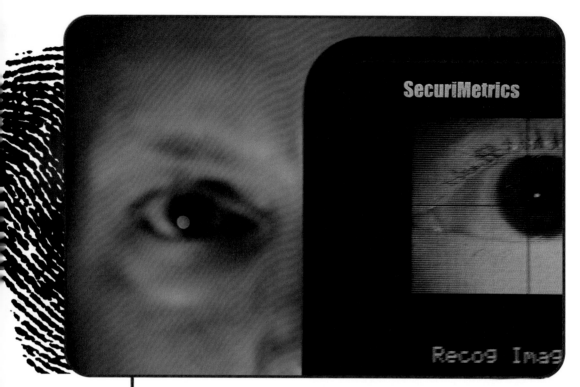

This person is getting his iris scanned. Iris scans are one of the many ways that biometric information can be harvested.

skin, hair, and eye color. Today, law enforcement officials are using sophisticated methods of identification that go far beyond what meets the eye. They are now also relying on biometrics, or the mathematical measurement of traits unique to each and every individual.

Fingerprints and DNA are two types of biometric information. Biometric information also includes an individual's handwriting and voice, and even the precise measurements of his or her face. Biometric identification techniques are growing more advanced

What the Future Holds

every year. Scientists are currently working on ways to identify people by their gait (the way they walk), the pattern of their retina, and even the unique way they type on a keyboard. Advances in biometric identification will make it easier for law enforcement officials to positively identify people. In turn, this will make it easier to find missing suspects and eventually to establish a suspect's guilt or innocence.

Fingerprint analysis has become more important as U.S. law enforcement has increased its counterterrorism efforts. People coming from other countries to work legally in the United States

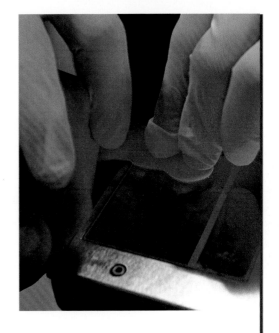

Fingerprint identification is helping to keep U.S. borders secure. Using IAFIS, an agent scans a person's fingerprints for comparison against the database.

are now fingerprinted before even being allowed into the country. These fingerprints can be checked against fingerprints of known international criminals and terrorists. Biometrics also play a role in keeping our borders safe. U.S. passports currently being issued contain a chip with the passport holder's biometric information on it. This will protect against passport theft and make it more difficult for known wrongdoers to escape the law.

Knowledge Is Power

New methods of collecting and analyzing evidence have advanced the field of forensics. But in the future, the most important changes to the field may involve the way evidence is accessed.

At one time, it was more difficult for different law enforcement agencies to communicate with each other. If a criminal left his or her fingerprints at a crime scene in Florida, for example, and then left fingerprints at a crime scene in Oregon, it could be difficult for police in both states to compare those prints. The Oregon police department would have no record of the prints on file in Florida and have no clue that their own mystery could be connected with another crime scene thousands of miles away. Today, electronic databases such as IAFIS contain the fingerprints of millions of people. Law officers can access a criminal's prints with the touch of a button.

The FBI is moving toward including biometric data in IAFIS, potentially creating a centralized database of hundreds of thousands of individuals that law enforcement officials around the world could tap into. Right now, fingerprints are the most common form of biometric information that law enforcement officials are likely to have on file and share with other police departments and investigative agencies. But what if there was a database that included not only millions of fingerprints, but also voice samples, DNA, and retina scans? The results of trace evidence analysis

Privacy

A massive and comprehensive collection of biometric information may help fight crime and terrorism, but it also raises a number of questions. Many law-abiding citizens feel uncomfortable with the notion that they could be so easily monitored, and they consider the collection of biometric information to be intrusive. Who will have access to this information? How might this technology be used outside the realm of law enforcement? Can the information it contains be passed on without your consent? Will this sort of technology lead to people being monitored without proper oversight? And what might happen if your biometric information fell into the wrong hands? Identity theft, or the theft of personal information for criminal purposes, is a growing problem in the United States. If identity thieves got a hold of someone's biometric information, they could possibly commit a wide range of crimes, some of which might be very serious indeed, making it appear as if you committed the crime.

As this kind of forensic evidence becomes more widespread, new laws regarding its application will undoubtedly be drafted to help protect law-abiding citizens from intrusion on their privacy by criminals and government officials alike.

CAREERS IN FINGERPRINT AND TRACE ANALYSIS

The owner of this card has registered personal and biometric information, such as her fingerprints and eye scans, with authorities. The card is issued as part of a federal test program that gives incentives, such as accelerated boarding time, to air travelers who participate in it.

would be filed right alongside developed fingerprints in a database of this sort. Should a centralized biometric database come into existence, it could completely transform the way law enforcement agents fight crime. The FBI is planning on updating IAFIS to include this additional identifying information. The new, integrated database will be known as Next Generation Identification.

In the rapidly developing world of forensics, there is no telling what the future may bring. There may be new methods of collecting and analyzing data, new ways to access that data, new ways to positively identify suspects, and new ways to ensure the innocent are not punished. Perhaps you will be a part of this evolution. It is never too early to start preparing yourself for an exciting and rewarding career in fingerprint and trace analysis.

Glossary

academic Someone devoted to teaching or conducting research into a particular discipline or field of study.

crime scene The physical location where illegal activity has taken place.

data Information.

DNA (deoxyribonucleic acid) Genetic information contained within the nucleus of the body's cells.

fingerprint The unique pattern of ridges and furrows in the skin of the finger.

forensics The use of scientific knowledge, methodology, and evidence gathering and analysis within a legal context.

internship A temporary, supervised period of employment. Internships provide people, usually students or recent graduates, with a chance to learn the career of their choice.

jury In a court of law, the individuals chosen to impartially render a verdict in a legal case.

laboratory A place in which scientific research or experiments are conducted.

stipend A (usually small) wage paid to interns.

surveillance The monitoring of people or objects.

terrorism Violent, illegal acts committed against civilians by individuals or radical organizations for political goals.

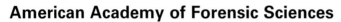

For More Information

American Academy of Forensic Sciences
410 North 21st Street
Colorado Springs, CO 80904
(719) 636-1100
Web site: http://www.aafs.org
A nonprofit professional society established in 1948 and devoted to the improvement, the administration, and the achievement of justice through the application of science to the processes of law.

California Criminalistics Institute
4949 Broadway, Room A104
Sacramento, CA 95820
Web site: http://www.cci.ca.gov
The California Criminalistics Institute, a unit of the California Department of Justice, Bureau of Forensic Services, provides specialized forensic science training to personnel who are practitioners in the field.

Canadian Society of Forensic Science
P.O. Box 37040
3332 McCarthy Road
Ottawa, ON, K1V 0W0
Canada
Web site: http://ww2.csfs.ca
A nonprofit professional organization incorporated to maintain professional standards and to promote the study and enhance the stature of forensic science.

For More Information

Federal Bureau of Investigation
J. Edgar Hoover Building
935 Pennsylvania Avenue NW
Washington, DC 20535-0001
(202) 324-3000
Web site: http://www.fbi.gov
The FBI's mission is to protect and defend the United States against terrorist and foreign intelligence threats, to uphold and enforce the criminal laws of the United States, and to provide leadership and criminal justice services to federal, state, municipal, and international agencies and partners.

International Association for Identification
2535 Pilot Knob Road, Suite 117
Mendota Heights, MN 55120-1120
(651) 681-8566
Web site: http://www.theiai.org
The IAI is committed to keeping persons in the forensic science profession informed, advancing the relevant sciences, and the dissemination of pertinent information through its publications.

Web Sites

Due to the changing nature of Internet links, Rosen Publishing has developed an online list of Web sites related to the subject of this book. This site is updated regularly. Please use this link to access the list:

http://www.rosenlinks.com/cif/fitr

For Further Reading

Ball, Jacqueline A. *Forensics*. Milwaukee, WI: Gareth Stevens Publishing, 2003.

Beaven, Colin. *Fingerprints: The Origins of Crime Detection and the Murder Case That Launched Forensic Science*. New York, NY: Hyperion, 2001.

Fridell, Ron. *Solving Crimes: Pioneers of Forensic Science*. New York, NY: Franklin Watts, 2000.

Genge, Ngaire E. *The Forensic Casebook: The Science of Crime Scene Investigation*. New York, NY: The Ballantine Publishing Group, 2002.

Lyle, Douglas P. *Forensics for Dummies*. Indianapolis, IN: Wiley Publishing, Inc., 2004.

Pentland, Peter, and Pennie Stoyles. *Forensic Science*. New York, NY: Chelsea House Publishers, 2002.

Rainis, Kenneth G. *Crime-Solving Science Projects: Forensic Science Experiments*. Berkeley Heights, NJ: Enslow, 2000.

Thornberg, Linda. *Cool Careers for Girls as Crime Solvers*. Manassas Park, VA: Impact Publications, 2001.

Bibliography

Anscombe, Nadya. "*Research Careers in Forensics.*" ScienceCareers.org. September 15, 2006. Retrieved June 2007 (http://sciencecareers.sciencemag.org/career_development/previous_issues/articles/2006_09_15/research_careers_in_forensics/(parent)/187).

Champod, Christine, Chris Lennard, Pierre Margot, and Milutin Stoilovic. *Fingerprints and Other Ridge Skin Impressions*. Boca Raton, FL: CRC Press, 2004.

Chisum, W. Jerry, and Brent E. Turvey. "Evidence Dynamics: Locard's Exchange Principle & Crime Reconstruction." *Journal of Behavior Profiling*, Vol. 1, No. 1, January 2000. Retrieved June 2007 (http://www.profiling.org/journal/vol1_no1/jbp_ed_january2000_1-1.html).

Cormier, Karen, Lisa Calandro, and Dennis Reeder. "Evolution of DNA Evidence for Crime Solving—A Judicial and Legislative History." *Forensics Magazine*, June/July, 2005. Retrieved June 2007 (http://www.forensicmag.com/articles.asp?pid=45).

Deedrick, Douglas W. "Hairs, Fibers, Crime, and Evidence." *Forensic Science Communications*, Vol. 2, No. 3, July 2000. Retrieved June 2007 (http://www.fbi.gov/hq/lab/fsc/backissu/july2000/deedric1.htm).

Dizzard, Wilson P. "FBI Plans Major Database Upgrade." *Government Computer News*. August 28, 2006. Retrieved June 2007 (http://www.gcn.com/print/25_26/41792-1.html).

Careers in Fingerprint and Trace Analysis

"Education and Training in Forensic Science: A Guide for Forensic Science Laboratories, Educational Institutions, and Students." National Institute of Justice. June 2004. Retrieved June 2007 (http://www.aafs.org/pdf/NIJReport.pdf).

Feige, David. "Printing Problems: The Inexact Science of Fingerprint Analysis." Slate.com. May 27, 2004. Retrieved June 2007 (http://www.slate.com/id/2101379).

Kanable, Rebecca. "Modern Forensic Science Today and Tomorrow." Officer.com. July 2005. Retrieved June 2007 (http://www.officer.com/publication/article.jsp?pubId=1&id=25192).

Lee, Henry C., and R. E. Gaensslen. *Advances in Fingerprint Technology.* 2nd ed. Boca Raton, FL: CRC Press, 2001.

McDougall, Paul. "Army Tries Matching Fingerprints to Catch Iraqi Insurgents." *InformationWeek.* February 13, 2006. Retrieved June 2007 (http://www.informationweek.com/story/showArticle.jhtml?articleID=179103427).

Scientific Working Group on Materials Analysis. "Forensic Paint Analysis and Comparison Guidelines." *Forensic Science Communications*, Vol. 1, No. 2, July 1999. Retrieved June 2007 (http://www.fbi.gov/hq/lab/fsc/backissu/july1999/painta.htm).

Simonite, Tom. "Fingerprints Reveal Clues to Suspects' Habits." NewScientistTech.com. April 3, 2006. Retrieved June 2007 (http://www.newscientisttech.com/article/dn8938).

Specter, Michael. "Do Fingerprints Lie? The Gold Standard of Forensic Science Is Now Being Challenged." *The New Yorker*, May 27, 2002.

Bibliography

The Technical Working Group on Crime Scene Investigation. "Crime Scene Investigation." U.S. Department of Justice. January 2000. Retrieved June 2007 (http://www.fbi.gov/hq/lab/fsc/backissu/april2000/twgcsi.pdf).

Trozzi, Timothy A., Rebecca L. Schwartz, and Mitchell L. Hollars. "Processing Guide for Developing Latent Prints." U.S. Department of Justice. 2000. Retrieved July 2007 (http://www.fbi.gov/hq/lab/fsc/backissu/jan2001/lpu.pdf).

Wade, Colleen, and Yvette E. Trozzi, eds. "Handbook of Forensic Services." Federal Bureau of Investigation Laboratory Division. 2003. Retrieved June 2007 (http://www.fbi.gov/hq/lab/handbook/forensics.pdf).

Index

B
background checks, 36
biometrics, 50–51, 52–54
   privacy and, 53

C
Counterterrorism and Forensic Science Research Unit (FBI), 42
crime scene investigators (CSIs), job duties of, 21–24, 27, 39–41

D
Dillinger, John, 11–14
disaster work, 44
DNA evidence/analysis, 17, 18, 27, 46, 48–49
dust evidence, 26

E
Eby, Margarette, 17
evidence collection toolkit, 40–41
evidence detection and analysis, in literature/on TV, 4–5
Evidence Response Team Unit (FBI), 43, 44

F
Faulds, Henry, 6, 10
Federal Bureau of Investigation (FBI), 6, 12, 14, 32, 35, 38, 42, 43, 44, 45, 52, 54
fiber evidence, 24, 25–26
fingerprint evidence/analysis
   contamination of, 21, 22–23, 39
   education needed for job in, 32–34
   equipment needed for, 40
   errors and, 18, 21
   future of, 46–54
   gathering of, 27–28, 39, 40
   history of, 4, 6, 9–11, 17
   importance of, 6, 8, 51
   known criminals and, 11, 14–15, 51
fingerprints
   attempt to remove, 12, 13
   patterns in, 29
   types of, 27
forensic science
   defined, 5
   future of, 46–54
   getting a job in, 32–38, 39–45, 48
   importance of, 8
   internships in, 34
Forensic Science Research and Training Center (FBI), 38
forensic scientists
   job duties of, 21, 30–31
   testifying in court, 31
friction ridges, 13
Fuhrman, Mark, 20

G
Galton, Francis, 6
glass evidence, 26
Gorton, Jeffrey, 17

H
hair evidence, 24–25, 27

Index

Honors Internship Program (FBI), 35
Hurricane Katrina, 44

I

Integrated Automatic Fingerprint Identification System (IAFIS), 17, 29, 41–42, 52, 54
internships, 34, 35

L

latent print examiners, 21, 22, 27, 30, 40, 44
latent prints, 15, 23, 27
Locard, Edmond, 6–7, 8
Locard Exchange Principle, 6–7, 46–47
Ludwig, Nancy, 17

N

Next Generation Identification, 54

P

paint evidence, 26
private investigators, 44–45

"Public Enemy #1," 12–14

R

Ramirez, Richard "Night Stalker," 15–16

S

Scotland Yard, 6, 9
Simpson, O. J., 18–20
Stratton Alfred, 9, 10

T

terrorism, fighting, 43–44, 51, 53
trace evidence
    analysis of, 24–27, 30
    collecting, 23–24, 39, 40
    contamination of, 21, 22–24, 39
    described, 5, 7–8, 24–27
    education needed for job in, 32–34
    equipment needed for, 40
    future of, 46–54
    importance of, 8, 18
tsunami, Southeast Asian (2004), 44

About the Author

The author of several educational titles, Jeffrey Spaulding grew up reading detective fiction. He currently resides in New York State, where he still peruses *The Complete Sherlock Holmes* when the mood strikes him.

Photo Credits

Cover, p. 1 © Richard Chung/Reuters/Corbis; p. 5 © CBS via Getty Images; p. 7 © Collection Roger-Viollet/The Image Works; p. 10 © Mary Evans Picture Library/The Image Works; pp. 12, 22, 47, 50, 51 © Getty Images; pp. 16, 19, 25, 31, 33, 34, 37, 45, 54 © AP Images; p. 28 © John Berry/Syracuse Newspapers/The Image Works; p. 38 © Anna Clopet/Corbis; p. 41 © www.istockphoto.com/Brandon Alms; p. 49 © SSPL/The Image Works.

Designer: Les Kanturek